# JOB INTERVIEW

## *QUESTIONS & ANSWERS*

**SANKET DESAI**
**MS, P.ENG, PMP**

Copyright ©2019 by TechCT Solutions, Inc.

All rights reserved. Without limiting the rights under copyright reserved above, no part of this book may be reproduced, stored in or introduced into a retrieval system, or transmitted, in any form, or by any means (electronic, mechanical, photocopying, recording, or otherwise) without the prior written permission of the copyright owner of this book.

# Table of Contents

INTRODUCTION ............................................................. 1

ABOUT THE AUTHOR ................................................. 5

PROJECT MANAGEMENT QUESTIONS ................... 8

SITUATION/SCENARIO-BASED QUESTIONS ... 104

SKILLS-BASED QUESTIONS ..................................... 126

INTERVIEW TIPS ......................................................... 144

FIRST FEW DAYS AS A PROJECT MANAGER .... 154

# Introduction

You only get one chance to make a good impression. Put your best foot forward with a winning Project Management interview preparation guide.

This short, comprehensive, easy-to-follow guide to winning the Project Manager interview aims to help you prepare yourself as the best candidate to stand out from the rest. Whether your goal is to acquire a new job as a Project Manager/ Project Coordinator or you are looking to switch your career as a Project Manager, this guide should help you prepare for your interview and acquire your dream job as a Project Manager in the first place.

Inside, you will find:

- Winning answers, tips, and techniques that will instantly attract the attention of employers, recruiters, and corporate headhunters

- Complete real-time scripted answers with no theory jargons
- Tips to help you sell your skills, brag about your attributes without sounding braggadocios, and detail your strengths so that they are more marketable and appealing to employers
- How to prepare for the interview start to end, designed specifically for the job you want
- Know what skills to specify and which to avoid
- And so much more!

The author, Sanket Desai, has compiled this book using his years of experience working in the field of Systems and Project Management. Sanket Desai has authored questions and answers using his real-life experience and his Project

Management's knowledge areas and process groups. In addition, Sanket Desai has received his Project Management Professional (PMP) certification from Project Management Institute (PMI) to enable him to understand the concepts of Project Management Body of Knowledge (PMBOK) and use those concepts while answering the questions included in this book.

This book comprises of Project Management interview questions that are highly expected in interviews. It does not include any domain-specific questions. This version of the book contains:

1. Project Management Questions
2. Situation/Scenario-Based Questions
3. Skills-Based Questions, and
4. Interview Tips
5. First Few Days as a Project Manager

Although you have no experience in Project Management but know the terminology of Project Management, this book is easily understood and offers examples, as well as visual representation, for readers to enjoy. This book hopes to increase readers' confidence and distinguish them from other applicants before the interview.

This book prepares readers with all the necessary tools and thought processes to answer questions in the interview confidently and increase the probability of standing out. The included tips and instructions will enhance the overall interview exercise.

# About the Author

Author **Sanket Desai** has been actively leading Project Management and Systems Engineering areas for over 12 years. An Alumni of Wayne State University, Sanket holds a Master's degree in Engineering, certifications in Project Management Professional (PMP) and Professional Engineer (P. Eng) from Ontario, Canada.

In addition to his many educational and professional accolades, Sanket has mentored and inspired many project management professionals and engineers all around the world. He has successfully led and managed numerous Project Management activities including developing project management plans, schedule management, resource planning, and change management in various domains.

His passion for his work has afforded him many opportunities, including the opportunity to thrive in his personal and professional endeavors. Now, he is expanding his reach to help other project managers to move forward in their careers by crafting the perfect interview guide for positions in this discipline.

# Project Management Questions

## 1. How is Program Management different from Project Management?

- Start by explaining the difference between both the terms, then focus on describing the difference between the two types as a Project Manager. It is not necessary to have a program management experience to answer this question. Take, for instance, if you are going for an interview for a Project Manager position, you will not be expected to have Program Management experience.

Sample Answer:

I have been involved in Project Management activities for several years, and I understand the differences between both management levels very well. None of these management levels are easy or difficult, but they have their own constraints and challenges.

A project is a set of activities carried out to deliver a product or a service. A program comprises of multiple projects that are related to each other.

As projects are smaller in magnitude and the objectives are smaller compared to a program, the management of a program and project is different in terms of the scope and the resources. Both management levels have their constraints and challenges. Due to the magnitude of a program, programs often involve multiple teams and may require more stakeholders as compared to a project.

A Program Manager needs to constantly work with all the Project Managers under that program to monitor and control the various aspects of a project, whereas a Project Manager needs to constantly monitor the various aspects of one project. A Program Manager is required to plan and allocate multiple projects and requires a significant effector in overseeing all the activities across the project within that program. A Project Manager manages resources, budget, and other management tasks of one project and reports to the Program Manager. A Project Manager often has to work with a more stringent budget and resources to achieve the goal as compared to a Program Manager.

In the end, both have challenges, but if the Project Management principles are executed appropriately, then both levels can be managed successfully.

## 2. How do you monitor and control a project against cost overrun?

- In this question, it is important to provide an example of your experience working on a project that faced cost overrun, but you should keep the answer on a high level without giving much information about the actual project and keeping the answer somewhat general. It is critical not to blame any individuals or the team for the cost overrun of the project you are going to discuss and keep it general.

Sample Answer:

I have worked on various projects in my past experiences, but one standout project was project

"ABC" and "XYZ" company. When I joined the company as a Project Manager, the project was going above the budget, and there were no measures in place to control this budget overrun. After my preliminary analysis, I realized that the main reason why the project was suffering was unclear project requirements. What was required was not clearly understood, and the work was being carried out without knowing what outcome will be accepted by the customer. Below are the steps I took to correct the situation.

- Clearly define and document the project requirement and gain the customer's agreement before any further work was executed. Even budget requirements became clear and helped us understand what the customer's expectations are.

- Factored in contingencies for different aspects of making sure the worst cases are planned for.

- Planned dedicated resources allocated to manage suppliers to eliminate miscommunication.

- Developed budget management plan with Earned Value Analysis (EVA) indicators to measure the project performance against.

- Frequently re-assessed and re-evaluated the project status to ensure budget management is constantly reviewed against any changes to the project and making sure the budget plan is re-aligned to the project goals.

Depending on the project, tools such as MS Project or Primavera can be used to monitor and control costs against planned budget. Budget is a critical aspect of any project, and constant stakeholder engagement is essential in bringing the team on the same page and making sure that stakeholders are aware of the on-goings of the budget and work towards achieving customer expectations.

## 3. How do you monitor and control a project against schedule overrun?

🎯 In this question, it is vital to provide an example of your experience working on a project that faced schedule overrun, but you should keep the answer on a high level without giving much information about the actual project and keeping the answer somewhat general. It is absolutely critical not to blame any individuals or the team for the cost overrun of the project you are going to discuss and keep it general.

Sample Answer:

"Schedule" is one characteristic of the project that can be planned for, but it needs constant monitoring because of continuously changing conditions in planned resources or customer expectations. I was managing one project "ABC" at the "XYZ" company, and the project suddenly faced changes in customer's expectations in terms of schedule. Till this point, the project was running in accordance with a planned schedule, but due to the change in customer's expectations, the project was behind schedule suddenly. The schedule planning was developed in a way to accommodate potential changes as well, but the proposed change, in this case, was significant and needed in the restructuring of the schedule planning.

As a Project Manager, my responsibilities included analyzing the change and incorporating the change request in the project once it has gone through change request processes. Once the change was officially accepted, I started the restructuring of the schedule estimations. I took the following steps:

- I started by analyzing changes against the current project status. The current status of the project was aligned with existing project status, but the new change introduced an amendment in the expected schedule.

- I analyzed the remaining tasks to be completed to achieve the target and started restructuring the priority of the tasks to meet the new expectation.

- Based on the new tasks' priorities, I started re-arranging allocated resources for each task to fast-track certain tasks.

- Also, wherever possible, I re-arranged the milestone dates to meet the ultimate schedule date.

- In the end, I involved all the stakeholders again to gain a consensus on the new schedule plan and ensure all the stakeholders' expectations are satisfied.

It is essential to understand that there is no way to eliminate all the problems that can cause schedule delays, but certainly planning of schedule can consider all the possible causes of delays and plan to manage potential changes.

Depending on the project, tools such as MS Project or Primavera can be used to monitor and control schedule against planned schedules. It is absolutely critical to incorporate contingencies in the planning of any aspects of a project.

## 4. How, as a Project Manager, can you manage changes to the project?

◎ The question is more for the process than the detailed example. Explain the change management process without giving details about a specific example of a change that you processed in your experience.

Sample Answer:

I believe change management is one of the most important aspects of any project. It is always advisable to remain flexible and open to accept changes. Changes can come from any stakeholders during any phase of the project. Any individual Project Manager should follow the pre-defined methodology/framework to follow the best practice. If the changes are accepted and incorporated efficiently, it increases the chances of getting positive feedback from the client and improves the chances of acceptance of the outcome.

I was part of a project "ABC" at the "XYZ" company where I developed the strategy to incorporate changes to the project. I developed a change management plan, which outlined the steps to manage changes throughout the project. Every change was evaluated against the impact and risk to the project. The change management process followed the following steps:

- Whenever a change is requested, the stakeholder fills out a change request form. This form includes details, such as change initiator name, change details, reason or justification for the change.

- This change goes through the Change Control Board (CCB), which evaluates the change, its risk, and its impact on the project.

- CCB is chaired by the Project Manager, involving all the major stakeholders to evaluate the impact.

- Depending on the analysis, the change is either accepted or rejected.

- If CCB accepts the change, then the Project Manager develops a strategy to implement that change.

- Eventually, that change is reviewed and verified to validate the implementation.

During this whole process, details of each step are documented to provide full traceability of added or removed features based on the change requested. Each change can be considered as a small project and can have stages, such as initiation, planning, execution, monitor and control, and closing.

## 5. What are the critical aspects you need to consider when planning for the project?

- This is a very broad question, and you do not have to explain multiple aspects. You should list some of the aspects you think are critical and then describe one as the most critical and then expand your answer around that one aspect.

Sample Answer:

I believe all the project aspects, such as initiation, planning, execution, monitoring & controlling, and closing are important, but planning is one which I think defines a project and plays an important part in project success.

I was part of a project "ABC" and the "XYZ" company where, as a Project Manager, I was responsible for planning the project. During the planning phase, I realized the importance of planning because it tells the stakeholders what you are trying to achieve and how you will be meeting the targets at the end of the project. It gives stakeholders the confidence that you have sold strategies and processes identified.

While planning, I felt the following aspects were very critical to a project:

- **Scope** – plan to develop and manage scope requirements through work breakdown structure throughout the project.

- **Stakeholders** – plan to identify and manage stakeholders to engage them early in the project to capture their requirements.

- **Schedule** – plan to develop and manage the project schedule aligned with project milestones.

- **Resource** – plan to identify and manage individuals or teams responsible for specific tasks according to their skills and availability.

- **Budget** – plan to outline budget details and constraints to meet the project requirement.

- **Risk** – plan to identify, manage, monitor, and mitigate project risks throughout the project life cycle.

- **Changes** – plan to anticipate and incorporate changes and backup plans for changes requested by the stakeholders.

- **Quality** – plan to identify the quality requirements of the project and strategy to achieve those requirements.

As I mentioned earlier, planning is the most critical aspect of any project, but it is crucial not to get carried away while planning and ensuring that realistic and feasible targets.

## 6. How do you improve your chance for customer acceptance?

- The focus of the answer should be the process and not blaming the customer for the rejection of any project or part of the project. The main reason a project outcome is rejected by the customer is that they were not informed during the development of the project and their inputs were not considered. So, if the customer's expectations are managed properly, then it improves the chances of acceptance.

Sample Answer:

The answer to this question involves stakeholder management as well. It is vital for any project to include stakeholders at all time, especially the customer as early as possible in the project so that they are aware of on-goings of the project and do not get any surprises at the end of the project.

I was a Project Manager at "ABC" company for "XYZ" project and, as a Project Manager, these steps can be followed to improve the chances of acceptance:

- Understanding customer requirements and acceptance criteria in the early stage of the project. Also, it helps to communicate your understanding of requirements and acceptance criteria with the customer to gain their approval early in the project.

- Ensuring the customer is fully entangled in the development of deliverables at various stages so that they are aware of the development and ensure no misalignment in expectations at the end.

- Take the acceptance criteria and incorporate them into developing acceptance testing procedures. This ensures you are testing the product against the acceptance requirement and whether the product is fulfilling these requirements.

- During the project, remain open to changes requested by the customer in order to align with the customer's expectations.

- Support customer's own acceptance verification and provide the information required of them during this phase.

In the end, this product/service was accepted by the customer. As a Project Manager, it is necessary to develop a strategy to outline acceptance planning at the beginning of the project by keeping the stakeholders updated.

## 7. Tell us about your experience in Agile development methodology.

◎ Describe the Agile methodology with advantages and disadvantages. Do not bring your preference in this answer and stay neutral and indicate that every methodology has its own strengths and weaknesses, but you have experience with many methodologies and you are comfortable with all.

Sample Answer:

I have been fortunate to be able to manage projects involving various development methodologies, such as waterfall and agile. These methodologies have their own advantage and disadvantage, and can be employed depending on the customer's and the project's needs.

I was leading a team as Project Manager for "ABC" company for the assigned "XYZ" project, where we used agile development methodology. Agile methodology is a methodology comprising project phases concept, requirements, design, implementation, verification, and deployment. In the Agile method, phases other than a concept and deployment are repeated multiple times to meet the customer expectation, which keeps the flexibility to incorporate changes. Each cycle of repeating requirements, design, implementation, and verification is called "sprint."

These sprints allow more opportunities to discover errors and constantly changing customer needs. Like every other development methodology, there are some advantages and disadvantages of Agile methodology.

The main advantage of Agile methodology is the flexibility to accommodate changes. This allows customers to alter their requirements as the project progresses. Also, it allows errors to be fixed in the next development cycle.

The main disadvantage of Agile methodology is that because it is flexible, it can suffer cost and budget overruns because of its changing plan.

Agile methodology is used when the scope is not defined clearly and the project is kept open for changes in the future. I am very confident that not only agile, but I will be able to manage other development methodologies as well.

8. **Tell us about your experience in the Waterfall methodology.**

   ◎ Describe the Waterfall methodology with advantages and disadvantages. Do not bring your preference in this answer and stay neutral. Also, indicate that every methodology has its own strengths and weaknesses, but you have experience with many methodologies and you are comfortable with all.

Sample Answer:

I have been fortunate to be able to manage projects involving various development methodologies, such as waterfall and agile. These methodologies cannot be compared as they are employed depending on the customer and the project needs.

I was leading a team as Project Manager for the "ABC" company for the assigned "XYZ" project, where we used waterfall development methodology. Waterfall methodology is a successive design phase, such as concept, requirements, design, implementation, verification, as well as operation and maintenance.

In the waterfall methodology, the next stage is started once one stage is completed. During the existing phase, it is not possible to return to either the previous stage or the next stage. Due to a lack of flexibility in the waterfall methodology, it is critical to plan each stage precisely and give enough significance to stakeholders' involvement to approve one stage to go to the next stage of the project.

There are several advantages as well as disadvantages of waterfall methodology. The main advantage of the waterfall method is the clear definition of the scope. The customer is clear on what they need, and there is no scope for changes. This allows better and clear planning of schedule and resources.

The main disadvantage of the waterfall methodology is that it is not flexible. The development stages can not be carried out in parallel fashion, and the customer sees the product at the end of the final phase. This limits the customer's inputs and increases the risk of the outcome not being accepted as there is no scope of incorporating the evolving needs of the customer.

Waterfall methodology can be considered as a best practice when;

- The customer is very clear about their needs/requirements
- There is more flexibility with resources and schedule.

9. **When was the last time you resolved a conflict successfully? What was the conflict and how did you resolve it?**

   🎯 In this question, providing an example is a must. Describe the process and then describe a situation where you resolved a conflict but do not take the side of who was right and who was wrong; stay neutral.

Sample Answer:

Project Management Body of Knowledge (PMBOK) from the Project Management Institute (PMI) talks about various conflict resolution methods, such as collaboration, confronting, forcing, withdrawing, etc. But I believe, as a Project Manager, you have to look for the most efficient way to resolve conflicts, which satisfies all the parties involved and move forward with the agreed-upon solution, which is best for the project.

I believe collaboration is the best method to resolve conflict. I can recall a conflict between my team and the supplier. According to the supplier, their product was meeting our requirements as described in the Project Agreement, but our team believed that the product is not compliant.

As a Project Manager, I set up a workshop with all the stakeholders, such as our team and the supplier. I went to the meeting with an open mind, listened to both sides' argument without being biased, and I treated both sides' opinions with respect. I did not want to avoid or delay the conflict, as it impacts the project. I had already planned for these kinds of situations when I started planning for the project. I asked my team to provide full traceability of our requirements in the Project Agreement. I assessed the situation and concluded that if the supplier does not meet the requirement, then the impact was not significant. So, when the supplier agreed that they are not compliant with the Project Agreement requirement, we provided them an option to either rework and meet the requirement or issue a change request through which we can change the requirement so that both parties can meet the requirement.

I think that the collaboration technique of conflict resolution has its advantages and disadvantages. It resolves conflicts in such a way that both parties feel achievement and the responsibilities are shared. The disadvantage is that this process may impose cost or budget overrun.

## 10. How do you identify and manage stakeholders of the project?

🎯 Stakeholders are anyone who has an interest in the project. Base your answer on the process rather than providing a specific example from your previous experience.

**Sample Answer:**

Stakeholders are one of the most important aspects of any project, which can define the project outcome. The process to identify and manage stakeholders is important for project performance. A stakeholder is anyone who has an interest in the project. To identify and manage stakeholders, the following steps should be taken into consideration:

- **Identify stakeholders** - stakeholders are identified through brainstorming, expert opinion, or past projects at the beginning of the project.

- **Categorize stakeholders** - analyze stakeholders' list and indemnity each stakeholder's interest and power in the project.

- **Manage stakeholders** - develop a stakeholder management plan, which will define processes to manage stakeholders and how to keep them informed.

- **Involve stakeholders** – it's important to involve stakeholders in every stage of the project to ensure their requirements are addressed and to improve the chances of acceptance.

- **Register stakeholders** - keep stakeholders' register with updated details as many stakeholders leave and new stakeholders join the project.

Project Manager understands the importance of stakeholders as the project success depends on stakeholders. A stakeholder management plan should be generated at the early stage of the project to ensure the processes are defined.

## 11. Have you used tools during your Project Management experience? What are they and how do they help in Project Management?

◎ Usage of tools varies project by project. You do not have to explain all the tools you have used, just pick the most common tools available and describe your experience with those tools.

Sample Answer:

I have used many Project Management tools in my past experiences, and they vary project by project. I can describe the common tools that I have used, and they are widely used for Project Management.

- **MS Project** – MS Project is one of the most used and important tools in Project Management. I have used MS Project to plan and manage schedule, including the activities and project milestones. I also used MS Project to assign budget and resources to tasks and tracked the progress to monitor the overruns.

- **IBM DOORs** – I used IBM DOORS for requirements management. DOORs provide functionalities such as requirements traceability and change management. Also, DOORs have configuration management capabilities and baselines the set of requirements that can be used by all the stakeholders.

- **MS Outlook** – I used MS Outlook as one of the communication tools for the stakeholders.

MS Outlook allowed sending/receiving email and setting up coordination meetings.

- **Skype/WebEx** – I used Skype/WebEx as another communication tool to set up conference calls for the stakeholders. When the stakeholders are not co-located, Skype/WebEx allows sharing screen and voice calls to ensure the impact of not having all the stakeholders at one location is minimized.

- **MS PowerPoint/Excel/Visio** – I used MS PowerPoint and Visio for analysis, such as Fault Tree Analysis, FMEA, and Work Breakdown. There are other software out there with the same capabilities, but depending on the project, the tools are selected.

12. What is WBS? How do you create WBS? How does it help in the project management?

🎯 Keep the answer to the process of WBS as it can be used for breaking down a task or deliverable. If possible, draw the WBS on a piece of paper or whiteboard (whichever available) as shown in **Figure 1** to demonstrate your understanding of WBS.

Sample Answer:

Work Breakdown Structure (WBS) is a visual representation of the hierarchical breakdown (see Figure 1) of the project work to be executed for projects or tasks by the team. WBS breaks down the scope of work in a manageable way and helps in planning for those tasks and subtasks. WBS can be used for:

- To breakdown for project deliverable or task.

- To breakdown task or deliverable to the lowest level until a single person or a team can be assigned to it.

- To allocate a specific budget to that task or deliverable.

- To apply a bottom-up approach to estimate budget, schedule, and resources for each task or deliverable.

- To identify the scope when it is not well defined, or when it cannot be broken down further.

- To trace cost or budget overrun related to a specific task or deliverable.

As a Project Manager, I have used WBS frequently in the projects I was involved in. I documented each WBS well and used a configuration management tool to ensure it is most up to date, incorporating all the changes to the project.

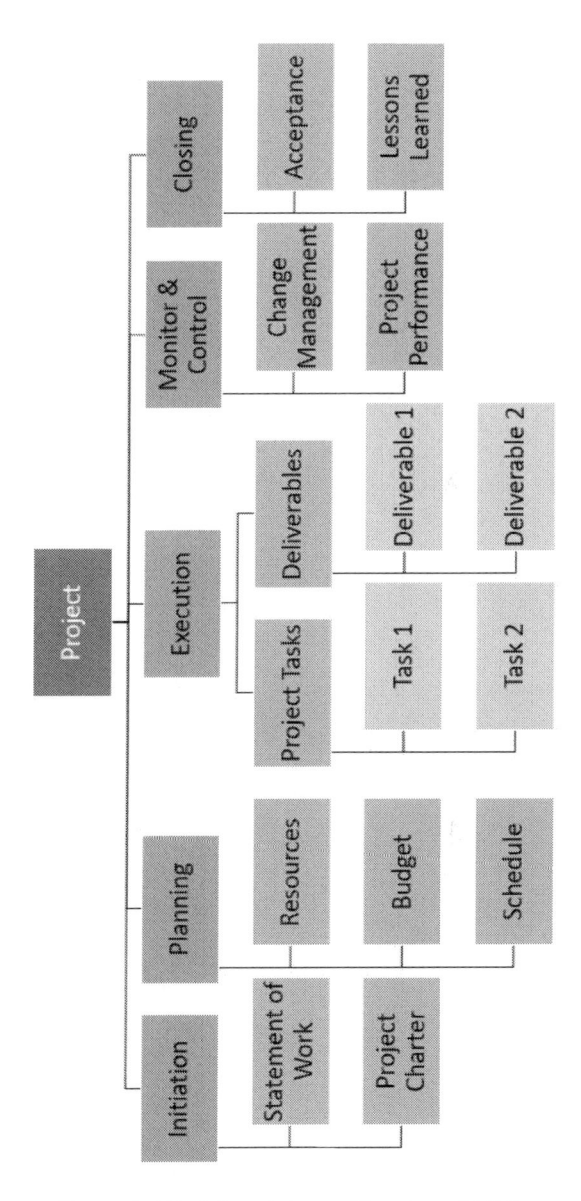

*Figure 1: Work Breakdown Structure (WBS)*

## 13. How do you manage the quality of a project? Describe some of the tools you used to manage the quality of the project.

- Quality is a broad subject. Keep the answer focused on the processes of planning, controlling, and assuring quality.

Sample Answer:

Quality is developing a product or process with minimal deviation from customers' expectations and with minimal deficiencies.

Quality management is vital as customers evaluate the quality of the product when they start using the product. Quality management involves continuous improvement during the development of the product, aligning the product with the customer's requirements and verifying the product continuously against the quality requirements in the contract.

Quality should be planned by identifying the quality requirements of the Project Agreement and the product should be developed in compliance with those requirements. Also, there are various techniques to monitor and control quality, such as Fishbone diagrams, design for six sigma, FMEA, fault tree analysis, and various charts. These tools, combined with management processes, can improve the quality of the product.

Quality assurance then verifies that the product is developed in accordance with customers' requirements and provides feedback to the requirements based on pass/fail criteria.

Quality is measured when the product starts getting used, and if it doesn't meet customers' expectations, then it may give a bad reputation to the project team. Also, as a Project Manager, one should always look to involve customers at every stage of the project if possible.

Providing customers with frequent updates and getting their approvals will improve the chance of the product being accepted. Other than the stakeholder's involvement, the Project Manager can also develop processes to test the product frequently to measure the quality of the product. Instead of finding the errors once the product is in operation, it is better to detect errors earlier and fix them. The cost of fixing errors is way higher in operation than fixing them early in the project.

## 14. How do you identify and mitigate project risks?

◎ Risk is a broad subject. Keep the answer focused on processes of risk identification and management.

Sample Answer:

Risk means the possibility of an event impacting the project performance. Risk management is planning to control and mitigate these events so that the project does not deviate from the expectations. As a Project Manager, it is obligatory to develop a Risks Management Plan from the beginning of the project. This plan should include the processes of identifying, managing, and mitigating risks for the project.

There are many tools and techniques for identifying risks, and some common ones are brainstorming, expert judgment, and lesson learned. Risks are identified best when all the stakeholders get to gather and brainstorm for all the possible project risks. A risk management expert and lesson learned from other projects are also effective ways of identifying risks to the project.

A risk register is a document that contains project risks-related information, such as ID, description, frequency, severity, owner, mitigation details, action plan, and status of the risks. This risk register is maintained throughout the project by keeping the stakeholders informed about any addition or removal of risks.

There are different ways to mitigate risk, such as accepting, eliminating, reducing, and transferring the risk. You can simply accept the risk and do nothing about it, but this method is not recommended. The best option is just to eliminate the risk completely if this is not possible, then risk impact can be reduced to minimal. The final option is to transfer the risk to either supplier, product, or develop operational procedures.

As a Project Manager, I feel risk management is one of the most important aspects of Project Management. If you plan well and are able to minimize the project risks, it significantly improves the chances of project outcome and success.

## 15. How do you manage contractors/suppliers of a project?

🎯 Supplier management is a broad subject. Keep the answer focused on processes of supplier identification and management.

Sample Answer:

Every project, at some point, involves suppliers. It is not possible to develop all the components in-house, and employing supplies can speed up the process. As a Project Manager, the following steps are required:

- **Buy Vs. Build Analysis:** you have to analyze first if you can meet the requirements in-house. Do you have the skills or the budget to develop project solutions in-house? If not, you have to consider procuring that solution from outside. This analysis is important, as it may impact the schedule and budget of the project.

- **Select Suppliers:** usually, suppliers are selected through the tender procurement process, where there is a separate procurement team that issues tender and shortlists the suppliers. This tender includes details like scope, evaluation, and acceptance criteria. Once suppliers and contractors are evaluated against this tender, they are shortlisted. Then after further meetings and workshops, the contract is awarded to one contractor.

- **Contract Type:** depending on the project, a contract type is established with the contractor. Contract types, such as time, material, and fixed-price contracts, are selected. Based on the negotiations between the procurement team and the contractor, a contractor type is selected, which is aligned with project expectations.

- **Manage Suppliers:** suppliers are managed throughout the project. A supplier management plan should outline the processes of supplier management. Interface with the supplier can be established through the contract. Frequent meetings, progress reports, and their requirements compliance reports are the tools to monitor the progress of the supplier.

- **Requirements:** The requirements management plan should outline the process to export the requirements to suppliers and suppliers can take these requirements and develop their own requirements. It is necessary to involve suppliers in the process of defining the requirements clearly and gain their agreement before the work starts. The project requirements provide bases for developing the acceptance criteria for the supplier.

- **Acceptance:** based on the acceptance criteria, the Project Manager can accept or reject the supplier's product or service. The details of the contract life cycle should be documented in a lesson learned register for future procurement tasks.

## 16. Do you have any experience with FMEA?

🎯 Remember that FMEA is a method for analyzing failure in a process or a product. Keep the answer to the purpose of the FMEA to the project. Give one example from your work experience to demonstrate how you developed the plan to manage FMEA for the project.

Sample Answer:

Failure Mode and Effect Analysis (FMEA) is a tool to identify and analyze possible failure modes for a product or service to establish the reliability and safety of that product or service. The FMEA related to design is called Design Failure Mode Effect Analysis (DFMEA), and the FMEA-related process is called Process Failure Mode Effect Analysis (PFMEA). FMEA can be used to analyze any risk that can significantly impact the project during any phase of the project. For each failure mode, FMEA identifies each possible cause that causes failure, severity, frequency, priority, and impact of that failure. FMEA also helps to identify the possible mitigation for the failures.

FMEA starts with picking a process or product that needs to be analyzed, and then failure modes are identified, which causes failures. Those failures are analyzed for their potential impact on the project. Eventually, the failure is categorized by priority based on its identified frequency, severity, causes, and detection methods of the failure. As a Project Manager, I have worked on several FMEAs and followed the following steps:

- Developed a plan which included the organizational structure and the review process of the FMEAs.

- Outlined the process of minimizing or eliminating a failure mode. Each failure mode should be mitigated by design first if possible, and then procedures should be developed to mitigate the failure mode further.

- Defined the process to verify that mitigation is implemented correctly once the mitigation is identified and implanted.

- Identified all the stakeholders that are impacted by the failure mode and obtain agreement on the implementation of the mitigation from the stakeholders.

## 17. Describe your experience with Root Cause Analysis.

🎯 Remember, Root Cause Analysis (RCA) is also a method for analyzing a problem or failure, but it is different from the FMEA. Keep the answer to the purpose of the RCA to the project. Give one example from your work experience to demonstrate how you developed the process of RCA for the project.

Sample Answer:

Root Cause Analysis (RCA) is a method of analyzing the causes of any specific problem.

RCA is different from FMEA. FMEA is predictive and used to analyze potential failures, while RCA is used to analyze failure causes once the failure has occurred. Similar to FMEA, RCA can be used at any stage of a project to analyze a problem and the causes.

RCA involves specifying symptoms and possible causes of a problem. It also outlines why the problem occurred and what can be done to avoid the re-occurrence of the problem. As a Project Manager, I developed a process to execute the RCA for one of the projects that I worked on:

- **Describe the problem:** describing a problem is critical because if it is not defined well, then the subsequent process of RCA can be inaccurate and can output an inaccurate analysis.

- **Gather information:** once the problem is acknowledged, collect all the data related to that problem, such as what happened, when it happened, why it happened, and any other information that will help to analyze the problem effectively.

- **Identify the root cause:** from the information collected, identify the root causes and factors that may have led to the problem. In this phase, causes are categorized based on their frequency and impact on the failures.

- **Mitigation:** either eliminate or minimize the impact and frequency of the problem. Mitigation can be implemented through design or operational procedures.

- **Verification of mitigations:** once the mitigations are identified and implemented, it is critical to verify them and see if they correct the problem as expected.

Just like any other tool, such as FMEA or FTA, it is essential to keep the stakeholder in the loop for

RCA as well. Stakeholders' agreement is required in order to implement the mitigations and other possible changes to the project due to RCA.

## 18. Describe your experience with Fault Tree Analysis.

🎯 Keep the answer to the process of Fault Tree Analysis as it can be used to identify the causes of a product or service failure. If possible, draw the Fault Tree Analysis on a piece of paper or whiteboard (whichever available) as shown in Figure 2 below. Provide failure example and its causes as an example of Fault Tree Analysis.

Sample Answer:

Fault Tree Analysis (FTA) is a systematic process and visual representation of identifying the root cause of failures in products or services. It is a top-down approach of classifying failure first and work down to all the possible reasons that can cause that failure. This process is used in the execution phase or the project to monitor and control the work being carried out to reduce and eliminate the failures.

As shown in Figure 2 below, the failure is listed at the top, and then the top-level causes 'Events A and B' are listed depending if both events cause the failure together or only one of them can cause the failure. If either Event A or Event B causes the same failure, then the OR gate is used to relate events to failure. If both Events A and B are required for failure, then AND gate is used. However, Events A and B are broken further until the lowest level of cause is identified, which leads to failure.

Providing one example of the FTA I carried out. We, as a team, were trying to figure out the possible causes of a process failure. The failure was that the contractor was not able to deliver the product on time. We identified two major Events that were causing this failure. Event A was that requirements exported to the contractor are clear, and Event B was that the contractor does not have enough resources to deliver the product on time. Event A was further analyzed to identify Event A1 and Event A2. Event A1 was that the requirements were not developed in detail for the contractor to understand. Event B2 was that the project changes were not incorporated frequently into the requirements. Eventually, we identified all the possible causes that were prompting the delay from the contractor to deliver the product. By eliminating or minimizing these causes, we could correct the situation and help the contractor to deliver the product on time and align the overall project schedule.

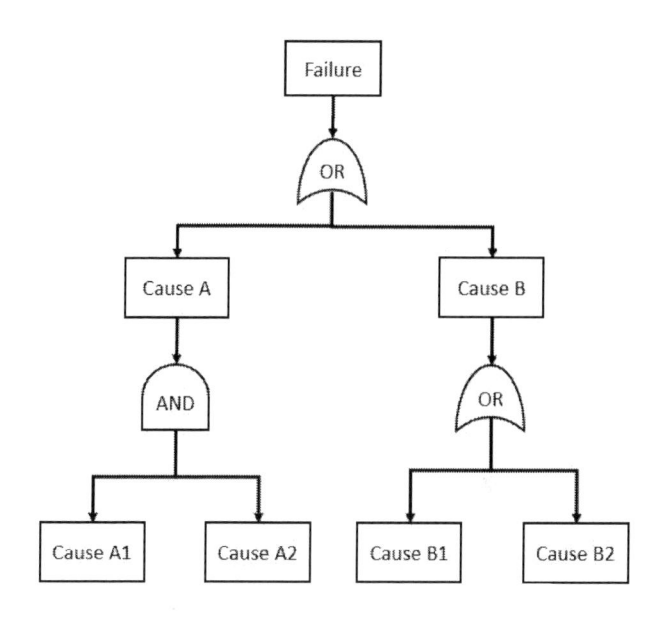

*Figure 2: Fault Tree Analysis (FTA)*

## 19. What is the difference between Critical Path and Critical Chain?

🎯 Explain each of the methods and then describe the advantages and disadvantages of both. Do not categorize one method as the best, but describe both methods and keep the answer open by stating that the selection of these methods is made based on the project requirements.

**Sample Answer:**

Critical path and critical chain are methods to develop or estimate the project schedule. While the critical path is task-based, critical chain focuses on the available resources.

The critical path is the longest sequence of activities for the project schedule that can be carried out without delaying the overall project. Each activity includes its own buffer. The critical path determines when each activity can be started earlier or later and finished earlier or later without delaying the project overall.

The critical chain method focuses on the resources compared to critical path. Critical chain method assumes that all the resources are available to perform the activities and adds a buffer for the overall project at the end.

There are pros and cons to both methods. Depending on the project requirements, either the critical path or critical chain method can be used for estimating the schedule. The critical path can identify the sequence of tasks, but there is a risk of having too much buffer and the project can suffer schedule overrun. The critical chain can help in estimating schedule without having a detailed understanding of the activities, but it assumes that all the resources are available for each activity. This can lead to an inaccurate estimate as resource availability can change at any time.

## 20. What is your understanding of Earned Value Analysis?

🎯 Earned Value Analysis is a broad subject. Keep the answer to the purpose of the Earned Value to the project and pick only a few critical parameters of Earned Value and elaborate on those parameters. Give one example from your work experience to demonstrate how you used Earned Value Analysis to identify schedule overrun issues and fixed it.

Sample Answer:

Earned Value Analysis (EVA) is a method for evaluating project progress and estimating time and cost at the completion date of the project. It compares the actual cost against estimated cost and actual schedule with the estimated schedule to provide the viewpoint of the project performance. Earned value calculations are used as one of the management tools to monitor the project progress and make an adjustment to align with the schedule and cost estimated at the completion of the project.

To support Earned Value Calculation, it is essential to break the project down into smaller parts and set the cost and schedule targets to compare them against Earned Value Calculation at that specific part. Depending on the Project Manager, some earned value parameters are selected to evaluate and report the project performance. The most common parameters are cost variance, schedule variance, planned value, and earned value. Cost and schedule variance demonstrates the difference between the actual cost/schedule against the planned cost/schedule. Planned value is the budgeted cost of the project work and can be calculated for different components of the project. Earned value is the used budget at a specific part of the project. These values can be used in combinations and with different calculations can help us demonstrate if the project is suffering from cost overrun.

A Project Management plan should precisely identify the parameters that can be used to monitor and report the project performance and outline a detailed strategy on adjusting the project performance if the cost and schedule exceed the expected value at any given point in the project. Also, the plan should include how the changes will be accounted for in the earned value analysis, As based on the performance report, the stakeholders with high interest & high power may alter their expectations with the project.

In one of the projects that I worked on in the past, I used earned value to report the project performance. Using MS Project, I performed an EVA for the project and determined that the project schedule was exceeding the regular schedule. I used Earned Value Analysis data to evaluate the situation and adjusted resources to specific tasks to ensure I achieve expected earned aligned with the planned value at the end of the project phase.

## 21. Why are well-written requirements essential for the project success?

◎ Requirements are the base of any project. It is essential to explain in detail why requirements are critical. Elaborate on the benefits that requirements provide to a project.

Sample Answer:

Well-written requirements provide a foundation for the project and support the project through the life cycle. The following are some of the significant benefits of requirements to the project:

- **Agreement** – requirements act as an agreement between the contractor and the suppliers. It establishes a common understanding between stakeholders. It provides the contractor an understanding of what the customer is looking for.

- **Reduces miscommunication** – require-ments eliminate the scope of misinterpretations and reduces miscommunication.

- **Reduces revisions** – requirements provide ample opportunities to correct errors at the early stages of the project. As the project progresses, it gets more expensive to fix the errors as the number of revisions increases.

- **Plan resources** – requirements are quantifiable of the work scope and help to estimate the schedule, budget, and resources.

- **Traceability** – requirements provide full traceability from contract requirements to any low-level requirements and, thus, make it easy to allocate and assign requirements to different stakeholders and trace it back to the contract requirements.

- **Verification** – requirements provide a basis for developing the test cases to verify requirements. Based on the outcome of the verification, requirements can be modified.

- **Acceptance** – requirements provide an understanding of acceptance criteria to contractors. They can be assured that if they comply with the requirements, the chances of the product being accepted will increase substantially.

- **Change management** – requirements support change management for the project. Any change to the project can be tracked and documented through the requirements with proper documentation.

22. What is your experience with require-

**ments development and management?**

🎯 This is a comprehensive question; you need to list the stages of the requirements development and management, as well as explain very briefly how the requirements should be collected. Then briefly explain each project phase that requirements go through. Focus more on the management where you manage changes, configuration, and verify the compliance of the requirements.

Sample Answer:

Requirements development is different if you are on a contractor side or the customer side. The typical requirements management chart is shown below in Figure 3. The customer requirements are very high-level and outline the high-level scope, performance targets, safety/security requirements, quality requirements, functional requirements, and acceptance criteria. The customer develops these requirements further and converts them into lower-level technical requirements. These requirements then can be used by designers, testers, and other stakeholders to comply with the Project Agreement. Requirements go through the following project phases:

- **Requirements** – requirements reflect the customer's needs, and they may include scope, budget, schedule, functional, operational, quality, and performance requirements. It may also have requirements derived from the high-level risk analysis. Customers develop these requirements based on their preliminary analysis, brainstorming, or expert opinion.

- **Integration** – integration is where the supplier puts the components together, making a system that satisfies the requirements provided by the customer.

- **Verification** – at this stage, various tests are carried out to ensure the product meets the requirements traced to these tests. Also, tests to validate the entire product are carried out to ensure it meets the customer's requirements based on pass/fail criteria.

- **Revenue/Operation and Maintenance** – At this stage, the product finally goes into operation mode and gets handed over to the customer in most of the cases, depending on the contract.

Requirements support the life cycle of the project, so it is vital to use an efficient tool to manage the requirements. IBM DOORs is a widely used tool, which manages the requirements through reliability, safety/security, verification/validation attributes for each requirement. DOORs provide the following capabilities:

- **Traceability** – IBM DOORs provide the feature of tracing the requirements from the project Agreement to the lowest requirements. This feature allows any stakeholder to trace the requirements allocated to them to contract and ensure they are not working on out of scope requirements, and each requirement they are allocated to is being developed further.

- **Configuration management** - an agreed-upon set of requirements is baselined frequently. The stakeholders agree to use baselined requirements until the next set of updates are made.

- **Change management** – once the set of requirements are developed and baselined, the next stage of requirements management can present changes, and these changes are reviewed through a change management plan and then incorporated into the requirements with proper documentation.

Requirements provide inputs to management plans, such as budget management, schedule management, and other management plans. Requirements management is a critical aspect of any project, and it can be modified as per the project needs.

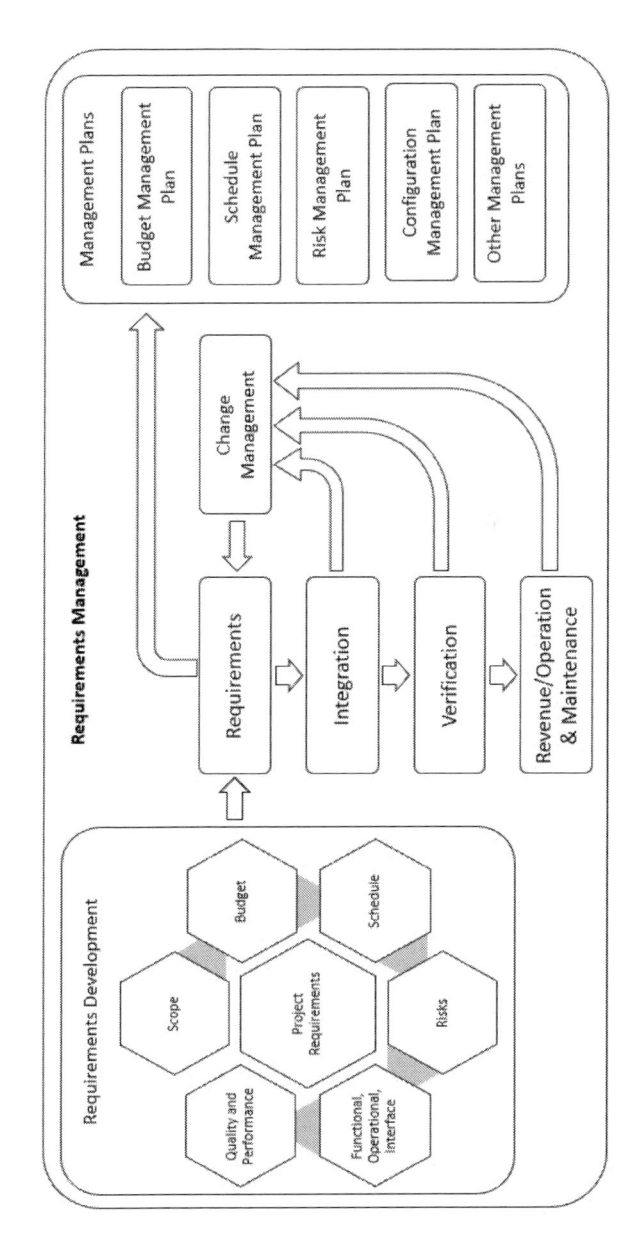

*Figure 3: Requirements Management Process*

## 23. How would you measure and manage the project performance?

🎯 Project performance uses Earned Value Analysis (EVA); however, the answer to this question should not be focused on EVA but should focus on project performance and how to measure, monitor, control, and report it.

Sample Answer:

Project performance is vital for communicating with stakeholders on how the project is going along. The project performance report allows stakeholders to make informed decisions in a way that may improve the outcomes of the project. Earned value is a tool that allows monitoring, controlling, and evaluating project against cost, schedule, and other variances. EVA should be performed periodically to analyze how the project is performing.

Project performance can be communicated through the project report, and the report should contain risks and their mitigation status, schedule, cost, quality, and safety performances. These performance parameters can be obtained from Earned Value Analysis, which evaluates all these parameters at a specific point of the project with the planned project values. Project performance specifies that the factors impacting performance are within the control or of the project team or not, which provides clear communications with stakeholders and their agreement on solutions.

Any variance in the performance of these parameters should be traced back to Work Breakdown Structure (WBS) to analyze where the deviation is coming from and identify the areas to improve. Overall, project performance gives perspective to the stakeholders on how the project is performing and provides high-level visual where adjustments can be made to re-align the project with expected performance.

## 24. What is the cost of quality concept?

- Do not blame the employees or the company for bad quality. Just state what the cost of quality is and explain how good and bad quality impacts the outcome of the project. Mention the advantages and disadvantages of spending appropriately on quality.

Sample Answer:

The cost of quality is the cost of achieving the project quality. There is a cost for bad quality as well as good quality. The cost of good quality is usually part of the plan, and it enhances the project outcomes. The cost of inadequate quality is unplanned and deviates the project from the planned performance targets.

There is a misconception about the cost of quality. If you spend appropriately on quality during the project development, then you will eventually save, make a profit, and earn a good reputation as the customer will be satisfied and there will be minimal re-work.

If you don't spend appropriately on the quality in the project phase, then you will save time and cost temporarily, but if the customer is not happy and their requirements are not met, then there will be a lot of re-work and cost of fixing errors. Also, you may lose future business from the customer.

As a Project Manager, you must assess the cost of quality for internal and external factors. The internal cost of quality is the cost before the product goes to the customer (revenue/operation). Internal quality can improve the chance of product being accepted by the customer. There will be less re-work and fixing errors. The external cost of quality is the cost after the product goes to the customer (revenue/operation). Bad external quality can result in repairs, claims, customer dissatisfaction, and loss of business.

## 25. What is your experience with Quality Assurance?

- Quality assurance is preventing quality issues before they occur. Do not get confused with quality control, which is about monitoring and detecting quality errors.

Sample Answer:

Quality Assurance is a process of taking steps to improve the product as well as the processes producing that product. Improving processes can eventually enhance product quality. Quality assurance is different from quality control, as quality assurance is focused on the prevention of errors, while quality control is focused on the detection of errors in quality. Quality assurance realizes the quality requirements and check the processes implemented against those requirements and ensures compliance with the requirements.

As a Project Manager, you should develop a quality management plan where you define all the quality measurement tools and performance indicators to be used to assure the quality. Quality is measured by evaluating the deliverables and verifying them against the requirements.

Once quality assurance is performed, change requests are generated to modify/update the process per findings. Eventually, project documents are updated to reflect the changes.

There are many tools to plan, monitor, and measure the quality. Tools such as cost-benefit analysis control charts, cause and effect diagrams, quality audits, and others identified by PMBOK. The primary tool used for quality assurance is a quality audit, which evaluates the existing processes per the requirements and recommends changes/updates to align the process with the expectations.

## 26. Describe the difference between Quality Assurance and Quality Control.

- This can turn out to be a tricky question as the interviewer wants to know your approach towards quality. Answer this question by describing both quality aspects, the difference between them, and picking the best one at the end of the project.

Sample Answer:

Quality assurance is about prediction and prevention of quality, whereas quality control is about checking and detecting quality issues. Quality is about the customer's satisfaction. If the customer is fully satisfied with the product once they start using it, then the product can be categorized as high quality. Quality assurance is always the preferred way of management as it decreases the cost of re-work and repairs at the end. Most of the projects do not put enough emphasis on quality assurance as it can be costly to invest in quality, but this cost can be easily recovered in the form of customer satisfaction, which leads to profits and future business. Quality assurance focuses on the process and rectifies them for a better outcome of the project, whereas quality control focuses on the outcomes and identifies errors and issues to make the outcome error-free.

Quality assurance and quality control are dependent on each other and work closely with each other. Quality assurance impacts quality control as the outcome of the project could be below expectations when the processes are not improved appropriately, and quality control has to put extra efforts in identifying issues. Quality control can find issues and provide input for quality assurance so that the processes can be corrected to eliminate specific errors.

Quality assurance utilizes tools such as quality audits to detect issues in the process and suggest corrective actions. Quality control uses tools such as control charts, cause & effect analysis, FEMA, etc. to detect errors and identify the gap between the actual and expected project outcomes.

Overall, as a good management practice, quality assurance should be a more preferred approach to detect issues early and prevent them instead of identifying them and fixing them as it may require much re-work and accrue cost and schedule over-run. Quality assurance could seem expensive in the beginning, but that investment can lead to customer satisfaction and earn a good reputation and future business with the customer.

## 27. Describe Human Resource Management.

🎯 This is one of the most important questions in the Project Manager interview as human resource is significant to the project, and if you manage them well, then it changes the output of the project. Answer this question in a way that you are not only focused on the project performance, but you also pay attention to the team's progress and wellbeing of the team members.

Sample Answer:

Human resource management is the management of human resources to the project by acquiring, managing, and releasing staff. As a Project Manager, a human resource management plan should be developed, outlining how and when staff will be acquired, how they will be managed, and how they will be released to the next project.

Acquiring staff includes specifying how the staff will be brought on board, depending on the project. Staff can be acquired through internal resources or external. Staff acquisition from external sources can be different from project to project, as it could be from the company's HR or staffing agencies depending on the project requirements.

Managing Staff includes assigning the roles and responsibilities, managing calendar, training, evaluating performance, and rewarding for the performance. The human resource management plan should specify performance indicators and reward details. The rewarding system can keep the staff motivated and impact the final outcome of the project.

Releasing staff includes planning for releasing the staff when the work is complete or assigning them to the next project. It is important to preserve skills and knowledge if possible, for further projects.

Overall, the human resource plan should include the process and activities to manage human resources for the project, including managing their health, safety, and security.

## 28. What is the "Law of Diminishing Returns?"

◎ Not all answers need to be descriptive. This answer should be a short and precise answer with one example from your work experience.

Sample Answer:

Law of diminishing returns indicates that at some point in the project, adding more to one parameter of the project pushes the output at a saturation point, and eventually, the output starts decreasing.

As a Project Manager, I realized that at one point, adding more resources to one particular activity does not necessarily improve or accelerate the completion of the tasks. I planned five resources for that activity, and then started adding more, the outcome was that more resources would need more ramp-up time, more time for collaboration which led to the issue of co-location, and more communication issues.

So, as a Project Manager, it is very critical to plan and allocate an appropriate amount of resources per activity; otherwise, more than required resources do not add any value to the outcome and even may decrease the value of the outcome.

## 29. What are the critical aspects of Project Initiation?

- Project initiation involves writing a project charter and usually prepared by the Project Sponsors. If you don't have experience writing a project charter as a Project Manager, then it is normal.

Sample Answer:

The project is defined on the basis of project initiation processes. If the project is set up accurately, the chances of a better outcome are higher. The project initiation phase comprises of the project charter and identifying stakeholders.

The project charter is usually not created by the Project Manager, so I don't have hands-on experience of generating project charter, but I am quite familiar with the document as the importance of the project initiation phase is known in the Project Management world.

The project charter is a document that describes the project overall, including the scope, the reason for the project, the processes to achieve the objectives, stakeholders' details, milestones, organization structure, identified risks, success criteria, as well as cost and schedule overview. The project charter authorizes the project and builds the case for the project.

The stakeholders' identification is very crucial as, in this phase, the project identifies the stakeholders with their level of support to the project. Not all stakeholders have the same level of interest in the project, so identifying their interest and power in the project helps the Project Manager to manage them efficiently.

As I mentioned earlier, if the project foundation is not strong, the structure may not last long. Solid project initiation is the foundation of any project, and a strong foundation gains more interest and support from stakeholders, which is ultimately the key to success.

## 30. Why is Configuration Management critical for the project?

◎ Do not focus on the configuration management tools, but rather elaborate on the management of the configuration of the project components. Describe how configuration management helps in achieving consistency and traceability in the project.

Sample Answer:

Configuration management is the process of sustaining the consistency for processes and deliverables throughout the project life cycle. Configuration management enables all the stakeholders to follow consistent procedures, supports any process-related audits if required, and accommodates changes to the project.

Configuration management includes the following aspects:

- **Planning for configuration management** - Project Manager should develop a configuration management plan and outline the configuration management processes such as stakeholders' engagement, baselining, control and monitoring configuration of project elements, change management, and status reporting structure and support internal and external audits.

- **Identification of configurable elements** - A document listing all the configurable project elements should be prepared. The input can be received from brainstorming sessions amongst stakeholders, experts' opinions, or past projects.

- **Control and monitor configuration** - the tools to control and monitor configurations should be identified, and the structure of the reporting configuration should be identified. The frequency of the configuration status reporting should be decided to all the stakeholders comply with the latest configurations.

- **Audit** - there could be internal or external audits for the project, and the configuration management should support the audits providing all the configuration data.

## 31. What are the critical aspects of Monitoring and controlling the project

**Work?**

🎯 Do not get confused between project performance and individual team member's performance. Focus your answer on how you plan to monitor and control the project work. The answer you provide should be general and applicable to any project.

**Sample Answer:**

Monitoring and controlling project work mean evaluating the planned project work against the actual project work. Monitoring and control project work involves continuously monitoring for changes. If any changes occur in scope, schedule, or cost, the Project Manager should have the plan to accept the change request and implement it in the project with minimal impact on the project overall.

The Project Manager should develop a Project Management Plan, which defines the processes and the methods to be used to monitor project work persistently and evaluate if it needs re-aligning to the planned work. The Project Management plan reflects the following aspects:

- **Look for changes/Detect changes** - Project Manager should frequently look for changes to the project by always involving stakeholders to verify if their needs have not changed and if they have, they are aware of the plan to incorporate those changes.

- **Analyze changes** – The Project Manager should also develop a change control team, which receives, analyze, and approves or rejects the changes. This team includes the applicable stakeholders. Based on the outcome of these processes, the changes are either accepted and implemented or rejected.

- **Implement changes** - The only changes that are incorporated into the project, which have gone through the change control team. The impact of the change, such as schedule, cost, or resources impact, should be documented and approved by all the applicable stakeholders.

- **Verify changes** – once the change is implemented, it has to be verified to ensure it is implemented as intended.

- **Reporting project performance** - project performance should be evaluated frequently to detect any deviation from the planned performance. Earned value is a widely used tool to evaluate and report project performance.

As a Project Manager, one should always be proactive and take preventive actions then corrective actions. The Project Management plan should plan for potential changes and add contingencies in the plan.

## 32. Describe a challenging project that you have worked on, and how you were able to manage?

🎯 It is critical not to blame any individuals or a team for the challenges faced by the project but talk about the challenges you faced during the initiation, planning, execution, monitoring, and control & closing phase to keep it general.

Sample Answer:

I have been fortunate to have got opportunities to work on many challenging projects, and I can recall one project specifically, where the teams were not co-located, and the communication was a big challenge. There are several other factors which impacted communications, such as different time zones, languages, and team expertise. When I joined the project, there were no set of communication protocols or processes that were established for all the teams involved. In a short time, I was able to recognize that because of the lack of communication between stakeholders, the needs/requirements were not being transferred properly, and deadlines were being missed frequently.

So, I started by developing the Communication Management Plan. This plan included the coordination details, such as frequency of meetings (weekly/bi-weekly), means of communication (skype, outlook, the meeting of minutes), communication protocols (language and two-way communication, etc.), identifying stakeholders and setting agenda for each meeting. Additionally, this plan identified a tool to log Meeting of Minutes, including action items and ways to track them by assigning responsible parties. By implementing this Communication Management Plan, a common understanding was established, and the coordination was carried out more efficiently and achieved desired results.

**33. Describe some of the most common issues with Project Management these days? What are the corrective actions?**

🎯 It is critical not to blame any individuals or a team for the issues faced by the project but talk about the issues raised during the initiation, planning, execution, monitoring, and controlling & closing phase to keep it general.

Sample Answer:

From my experiences, I believe that the two primary challenges are stakeholder management and incorporating lessons learned from similar projects. It is critical to emphasize these two aspects right from the beginning of any project. In the early stages, it is important to identify all the project stakeholders, their needs, and manage them based on their interest in the project. For example, in one project, not all the stakeholders were involved at an early stage, and later on, it was discovered that that stakeholder's requirements were not captured accurately, and it did not align with the overall project scope. This led to multiple revisions and cost and budget overrun.

It is critical for any project to have a stakeholder management plan, identifying the processes to identify and engaging all the stakeholders for the project. Also, an equally important aspect is the lesson learned. Ideally, at the end of every project, the manager should document lessons learned according to industry practice, so the same mistakes are not repeated in the next project. When a project is in the initiation and planning stage, a well-documented lesson learned scenarios ensure that the same errors are not carried forward and provide opportunities to plan for those potential errors by analyzing them and mitigating them as early as possible.

## 34. Why do projects fail?

- It is critical not to blame any individuals or a team for project failure but talk about the common reasons due to which projects fail. Also, provide an example of a project which you were not part of and say that through your research and colleagues, you have learned about these common reasons for project failures.

**Sample Answer:**

Projects fail due to various reasons. Also, following the Project Management process (process groups and knowledge areas) does not guarantee success, but if applied correctly, it improves the chances of success. It is vital to outline the requirements of the project (work scope definition) clearly. The project requirements specify roles and responsibilities, and it acts as an agreement between stakeholders. It establishes a common understanding between parties and reduces the chances of miscommunication. In most of the projects, the ambiguous requirements lead to different interpretations and form confusion.

The other aspect which leads to project failure is project integration. Project integration is the process which takes all the processes of the project into account and makes them work together. The Project Manager needs to manage and control project parameters such as scope, schedule, cost, risks, changes, stakeholders and resources, and costly analyze the trade-offs. If one parameter is performing below expectations, then other parameters can be adjusted to compensate for the damage. This is easier said than done, but as a seasoned Project Manager, one needs to identify a tool to manage these project parameters to achieve the desired results consistently.

# Situation/Scenario-Based Questions

Situational interview questions provide an excellent way for an applicant to emphasize his/her past accomplishments and highlight outstanding professional skills and competencies. These questions also provide ample personality assessment criteria' for the Hiring Manager/Human Resources panel to make hiring decisions.

Not all questions in a job interview are situational; however, an upward trend is being increasingly noticed with the number of such questions being asked (consider at least 1/4th of interview questions). It is recommended that you sketch high-level situations related to the few topics below (and more!) before the interview day and practice your speech. Unlike, other typical 'content-oriented' or 'technical' questions, these should sound less like a rehearsed speech (don't be too artificial) but should not make you go blank (or spell-bound). If you have a few situations crafted/thought-through from previous experiences in your head, chances are you can readily employ those situations and swiftly navigate through such questions. Your answers should emphasize that you focus on issues and facts as opposed to people's or your own opinions.

Here is the most known and powerful STAR acronym for answering these questions:

**S – Situation:**

What is the context (background/ scenario) of your situation?

**T – Task:**

What impact did the situation have on your tasks? How does the situation affect you?

**A – Action:**

What is it that you did? How did you resolve the issue? What strategies/skills did you employ?

**R – Result:**

What was the outcome of your actions? Did you learn something new about yourself? Learned new people/task management skills? How does this improve you today? How can you bring this new strength to the table for the job at hand?

We like to think of these questions as 'story-narrations' with a recount of positive learning for you. We recommend not to overemphasize 'What Happened?', instead, emphasize 'What did you do?', all in no more than three minutes. Oftentimes, candidates focus more on following the exact STAR pattern to answer such questions, which should not be the case. If you cover the high-level problem in a situation and its resolution, you are on the right track.

1. **How do you resolve stakeholders' conflicts?**

    *OR*

    **What if the stakeholders do not agree with each other's solution or have an internal conflict? How would you manage such a scenario?**

    🎯 Situations or conflicts may arise due to different reasons. Sometimes different stakeholders may have varying understandings of a single component of a system, or simply put, some requirements are complex in nature. No matter what the situation is, exceptional coping and coordination skills can mitigate any conflicting scenarios. Discuss having experience in resolving stakeholder conflicts. It is not always necessary that you have experienced the same situation from before; in that case, you can talk about any related or similar situation from the past.

Sample Answer:

As a project manager, you should always be prepared to resolve conflicts that may arise during the project. A project involves many stakeholders, and as the number of stakeholders increases, more conflicts are inevitable. Some of the most common types of conflicts in the project are introduced by communications gaps.

I remember when I was working on project ABC in company XYZ, where the contractor and my team had a difference of opinion for the scope allocation. The project agreement (contract requirements) required multiple contractors working together for final systems integration. Contactor A was supposed to provide a system, and contractor B was supposed to provide the infrastructure required to install that system. There was some aspect of the scope that was overlapping, and the conflict raised regarding who is supposed to provide what information.

I used my analytical and problem-solving skills to evaluate the entire situation and decided to gather information separately from both groups. This approach allowed me to focus on both groups' rationale for differing recommendations/requirements.

After my analysis, I realized that the project agreement requirement was indeed not clearly defined and required a change in order to clarify the situation. When I started the project, I already planned for such situations through my project management plan. I prepared a change request and asked both contractors to submit their cost and schedule estimates to provide the required scope. I received the estimates from both contractors and evaluated those estimates against project needs. After my thorough analysis, I organized a meeting with all the stakeholders and presented my analysis to pick one contractor who can meet all the requirements of the project with minimum impact to the project. After the stakeholder's agreement, I awarded the work to a suitable contractor.

By channeling effective stakeholder management skills, I was able to analyze the situation, resolve conflict, and create an amicable work environment.

From all my past experiences, I can conclude that conflicts are inevitable in any project, but it is their thoughtful management that can harmonize relationships without risking the timely implementation of solutions.

2. **Do you have experience in managing stakeholders who may change requirements frequently?**

   - Respond to this question in affirmation by developing a story (STAR technique of answering). The intent of this question is for the interviewer to figure out whether you can manage a project with the presence of 'various stakeholders with various needs.' Express yourself in a way that reinstates that you have 'people' as well as 'project management' skills. Talk about the importance of staying on track with the project timeline and how you have achieved that by following a methodical strategy when challenged by stakeholders with uncertain requirements.

Sample Answer:

In my last project, I worked with a team of advisors, most of whom joined the project midway through the Requirements Gathering phase. As such, they did not possess the same understanding of the project as others who were on board from the inception.

As quite predicted, I had the struggle of bringing them on the same page as everyone else when running requirements workshops like focus groups and interviews. Often, I was given a set of requirements by the stakeholders, which were either already implemented or deemed out of scope.

When this started affecting the project timelines and deliverables, I had to rejig my requirements elicitation approach. I started to kick-off my meetings by going over the project scope document with the stakeholders and reiterate the objective that we were set to achieve.

This provided a conscious undertone to our activities as well as set the direction for requirements elicitation. I also employed the reverse-engineering technique, which at its core, endeavored to link back all the requirements to the primary objective of the project. Although this approach worked the majority of the times, yet in certain situations where the requirements did not match the stakeholders' expectations, I developed white papers (problem description, my understanding, and a possible solution), which eliminated redundant/vague requirements.

I also have vast experience with change management/change control process in various methodologies. In situations where stakeholders made modifications to requirements, a formal, written-out change order form was expected to be filled out by them.

I thoroughly assessed change orders to ensure new/revised requirements did not lead to scope creep or budget/resource issues. A review committee was set up to analyze the impact of those changes on the project's cost and timeline. All these measures ensured requirements did not just 'wishfully' change, but in fact, were vetted through a formal and coordinated process.

3. **Describe a time when you introduced a new idea or process to a project and how it improved the situation?**

   ◎ This question is centered on knowing your interpersonal and decisive analytical/ leadership skills more than knowing which organization/ position you were in at the time you suggested a new idea. This question can be answered in multiple ways. You can either discuss a novel approach you took to something that was stagnated/ inefficient/outdated, either in the work sphere or from personal experience. The question is aimed at knowing if you harbor innovation/creativity and can be a self-starter. Your answer should convey that you can identify an issue that needs improvement and act on it immediately to bring in heightened efficiencies.

Sample Answer:

Implementing new ideas to improve project performance is one of the obligations of a project manager. A proactive manager continually looks for opportunities to improve project outcomes by using management skills and implementing new ideas.

I remember when I was working on project ABC in company XYZ, where I joined the project midway. The project was having some serious issues as the project was not progressing, as the stakeholders expected.

There were a number of workshops held between us, the customer, and the contractors to resolve design-related misinterpretations. I analyzed the situation and realized that there were communications gaps between two parties, and the same topics were discussed in the meeting again and again and costing the project.

I started by brainstorming to correct the situation. I developed the process of briefing papers; in these documents the contractor provides their understanding of the design requirements, and we, as the customer, review their understanding and approve so they can implement that agreement in the design. This process was compliant with the configuration management plan, which allowed to generate a document number of each briefing paper and helped both parties to trace and manage those design agreements.

I updated the communication management plan and system integration plan reflecting the new process. This new process reduced miscommunication and re-work significantly.

4. **Describe a time/situation when you worked under a high-stress environment. How did you handle the high work-pressure and your workload?**

   - This question is asked because the hiring manager is interested to know how you manage piled-up work. It is common for project manager positions that you would be required to take on multiple against tight deadlines. Your answer should directly clue the employer into thinking that your organization skills are an excellent match for the position. It is also recommended to not 'pass on the blame' to others in your team, as it may look bad on you as a team player. Talk about the 'situational-reasons' for the overload and not 'personal'.

**Sample Answer:**

Project managers often face pressure to meet the project expectations as schedule, cost, and resources requirements are always changing. I will say planning plays a significant role in avoiding or mitigating these pressure situations.

When I started working in this company as a project manager, I ensured to spend enough time and effort to develop a project management plan that can help manage the project efficiently and avoid pressure situations. My team was the contractor in this project, and the customer was constantly changing project requirements, which was adding the constant pressure to adjust resources to achieve frequently changing project milestones.

While developing the project management plan, I paid extra attention to stakeholder management, change management, and resource planning. As these changes were coming in, I was constantly working to adjust my resources to meet deadlines by using tools such as critical path, critical chain, and earned value analysis to ensure that sufficient resources are allocated to tasks. I increased the frequency of stakeholder engagement to ensure the project progress is reported, and new requirements are captured as early as possible to reduce the amount of re-work. Also, frequent stakeholder engagement allowed us to identify the task priority and adjust our resources according to those priorities to achieve the desired project outcome.

My change management processes allowed us to ensure the new requests we are receiving from the customer are properly documented, and the required resources are allocated to analyze, implement, and verify these changes.

Careful planning and organization of tasks helped deliver the project within the allotted time. I learned how to effectively manage project teams and work as well as not let the high-stress environment affect my performance.

5. **You are asked to lead a complex project with multiple stakeholders involved. How would you ensure proper stakeholder management and reduce communications gaps?**

   ◎ The answer to this question should establish your experience and know-how of conducting different types of stakeholders. The employer wishes to learn what would be your approach to managing stakeholders with different levels of interest and power in the project. In your answer, mention how your stakeholder management skills can help the project.

**Sample Answer:**

If the stakeholders are not managed effectively, then the project outcome can be significantly get impacted and may accrue cost and schedule overrun. In one of my projects, I served as project manager/integrator and worked with multiple stakeholders to integrate the system. For the project, I was required to work with several contractors and vendors.

I encountered challenges when stakeholders try to assert domination in the project. This was exerted by the high influence they had on the overall solution when compared to other contractors relatively smaller in experience and sizes. Nevertheless, each contractor had a significant role and input to provide in finalization of the project, and I made sure to not to compromise that equation.

Initially, I worked with the representatives of contractors directly in the meetings. As consultations progressed, more stakeholders joined the projects, and the discussions started getting more complex, and the newer/smaller stakeholders felt their input hardly get counted in meetings. This, in turn, resulted slow progress of these discussions, and budget and schedule started getting impacted.

To resolve the situation, I started working with each stakeholder on a one-on-one basis. Before each meeting, I would create a presentation by extensive document analysis and set a clear agenda for the meeting. This approach provided a better solution to the stakeholders' needs, regardless of their experience and organization size, opportunities to present their view on the project, and increased their participation.

Through my tact and judgment, I was quick to change gears on my stakeholder management without derailing the schedule from where recovery would have cost a fortune.

# Skills-Based Questions

1. **What are your strengths?**

    🎯 Always discuss your Project Management expertise or related strengths. Strengths from other areas could be brought up as well, with greater emphasis to be given to highlight your forte in Project Management. Remember to prepare at least 2-3 strengths ahead of the interview. Generally, strengths do not require an explanation as compared to weaknesses.

**Sample Answer I:**

While working as a Project Manager for last X

years, I have sharpened my analytical thinking, problem-solving skills along with my competent Project Management skills such as time, cost, and resource management. This has allowed me to manage projects even under tight timelines and limited resources.

Sample Answer II:

I thrive on new challenges and look for avenues to gain as much knowledge as I can. I thoroughly enjoy reading project management articles and connecting with industry experts to gain from their insights.

2. **What are your weaknesses?**

    ◎ You may choose to talk about your weaknesses from other or related technical areas, but it is recommended not to bring forward vulnerabilities from the Project Management profession. It has been noted that employers/HRs do not appreciate 'disguised weaknesses' (attributes which truly are strengths, but one chooses to position them as weaknesses), such as:

- I am too good to be true, and I think my team members have an issue with that, so I am trying to counterbalance my skills.
- I like to make sure that I achieve surpassing perfection in my work; thus, many times, I find myself spending way extra time ensuring my work is error-free.
- When I'm working on a project, I don't want just to meet deadlines. Rather, I prefer to complete the project well ahead of schedule.
- Being the perfectionist that I am, I

do not just meet the deadlines; instead, I complete all my tasks way ahead of time as compared to my counterparts.
- These will be professional faux-pas, so stay away from sugar-coated weaknesses!

Sample Answer I:

Sometimes, I spend more time than necessary on a single task or take on tasks personally that could easily be assigned to someone else. Although I strive to not to miss a deadline, it is still an effort for me to know when to move on to the next task. To overcome this, I have started assigning priorities to my tasks and organize project deliverables accordingly.

Sample Answer II:

I struggle with delegating tasks to team members, due to trust and performance reason that they may not be able to accomplish the tasks within the given timeframe and with the same quality. To get rid of this weakness, I have started to define milestones and deliverables before handling work to anyone. This approach has hugely benefitted me and others in my team as trusting working-relationships are building.

Sample Answer III:

I have great experience of working on multiple tasks concurrently, but there are some tasks which require extensive research, analysis, and undivided attention. To deliver all wide-ranging tasks within the expected timeline, often, I have observed that quality gets compromised. To overcome such, I now ask for help early-on or request extensions so that those tasks requiring expertise and extra due-diligence, do not get compromised.

Sample Answer IV:

When responding to stakeholders' requests/

requirements, I used to spend a lot of time discussing simple requirements, which had led to failure in accomplishing meeting agenda in a single workshop. I have learned to prepare for a solution before the session, to eliminate extended discussions over simple project needs.

## 3. How do you handle failure as a Project Manager?

🎯 The way you answer this question could stand you out in terms of how you handle and resolve project or requirements failure. This question is disguised under two intentions:

**I.** How transparent are you in acknowledging your failures?

**II.** What have you learned, or what is your approach when you are struck by a failure?

You do not necessarily need to talk about project failure, but if you choose to, common examples can be - not meeting the timeline, not meeting the client's exact expectation. Discuss your many skills such as organization, people and, task management skills, your proficiency in improving tasks and processes. You should sound honest, and convincing at admitting your failure(s) while building the interviewer's confidence that with your capabilities (e.g., keeping sanity intact, focus, and positive attitude), you can overcome any project failure.

**Sample Answer:**

I consider even a small issue in the project/not meeting the client's requirement on time as a failure. I like to be professional and proactive in terms of my work commitments, regardless of the methodologies and processes being followed. This strategy allows me to stay on top of my tasks and strike a balance between project scope and timeline without getting failures affect my performance for long.

In my last project, we had missed the deliverable milestone date by two weeks, which was mainly attributed to sudden personal issues faced by three of the key members in my team. I, being the Project Manager, did not plan for three team members having personal issues at the same time. As anyone going through a tough time may do, the performances of my team members started deteriorating. My mistake being, I failed to plan and flag this situation to senior management early-on and not devising a mitigation plan. Nevertheless, as we missed the 1st schedule date, after that, I was quick enough to realize where we had gone wrong and asked for additional resources. Looking back, I can say, better resource planning and asking for timely help were my two biggest learnings. I take pride in knowing that a calm and composed head and determination to strike back on, with improved zeal, are my mantras to handle any failure and for it to not become a repeated activity.

4. **Where do you see yourself in the next 5 years?**
   **OR**
   **What is your career plan?**

   - If you are just starting as Project Manager, focus your answer more on building Project Management skills. Saying something like, "I want to become a Senior Project Manager or a Program Manager, right away" may not be a great strategy.

     Say something like you envision becoming a Program Manager in the long run (after 3 years or so). Mentioning of quick/immediate transition does not show long term affection/connectivity/ commitment to the Project Management job. If you are an intermediate/senior Project Manager, then discuss how you want to avail experience in different domains, roles, and your commitment to achieve that. You can bring forward any certification you are preparing for or any efforts made to enhance your skills/education portfolio.

**Sample Answer:**

I've been practicing Project Management for the last 2 years, and I want to grow as a Project Manager and eventually a Program Manager. To attain those roles, I am sincerely expanding my domain know-how and learning the duties/responsibilities undertaken by a Project

Manager.

I just completed my Project Management Professional (PMP) Certification, and I am now preparing for Master Project Manager (MPM) certification. I relentlessly look to brush-up my project management skills time to time.

## 5. What are the skills required by a Project Manager?

- It is recommended to discuss the few skills listed in the first point below. In addition to that, you can (should) discuss other Project Management skills. Having stated positive skills does not require in-depth clarification. Simply stating them and/or elaborating in 1-2 sentences, is good enough.

Sample Answer:

The most important skills required by a Project

Manager are stakeholder management and requirements management as they can significantly impact the outcome of the project and can contribute to the success of failure of the project.

I have learned and developed different other Project Management skills like risk management, quality assurance, and quality checking. I like to stay abreast with new tools and techniques as I thrive on learning new technologies.

6. **Why did you leave your last job?**

    🎯 It is not recommended to talk about personal situations at the interview time. (e.g., conflict with team/manager).

Sample Answer I:

My recent project is over.

Sample Answer II:

Due to the simple and repetitive nature of projects, there is no opportunity to grow in my current job, and I want to further pursue my career as a Senior Project Manager, and eventually Program Manager, by facing newer challenges and expand my skill-set.

Sample Answer III:

I have worked at my current organization for the last four years and have honed my experience and skill-set. I feel confident to take on more responsible roles, which will steer me to think/act outside of my comfort zone.

Sample Answer IV:

My company is downsizing currently, although, I have not been notified of any decisions, I am proactive in my job search to secure a position in an established and continuously growing organization, like yours. My Director is well-aware of the situation, and I will be able to provide excellent references, if and when required.

7. **Why are you a good fit for this job?**
   **OR**
   **Why should we hire you?**

   🎯 You should match the job requirements, including education, certification, and experience, while discussing this question. If you can match the 'must-have skills' listed in the posting, it significantly improves your chances of landing the job offer. If you are pursuing something (educational courses, honing a skill), you should mention that as well. Extra points can be accumulated if you can convince the hiring manager of how you would be an asset to the company.

In your answer, you should not sound too boastful/bragging about your experience, which may imply 'employer's loss' in not hiring you.

**Sample Answer:**

As your company is looking for a candidate with decent experience in using project management tools, I have been using project management tools such as MS Project, Primavera, and IBM DOORs in the last X years. In particular, I have experience setting up a complete project management system from scratch.

This position's educational requirements also assert a preference over someone with a bachelor's degree. I earned my bachelor's in Computer Science and went on to completing master's from XYZ University with a special area of emphasis in project management. I also have PMP certification from PMI.

I consider myself experienced in other

aspects of project management such as cost, schedule, resource, and risk management of the project, including analytic tools such as critical path, critical chain, and earned value analysis.

In addition to bringing superior communication and proficient documentation skills to the table, I bring along the fine art of building and maintaining client relationships.

# Interview Tips

- Dress professional and sharp. Go with tried and tested clothing ensembles:

  **For men:** Dark color suit, a pair of dress shoes, a matching belt with shoe color

  **For women**: A suit or dress with a blazer can also be opted, depending on comfort

- Pay attention to small details in your attire: tied buttons for shirt/sleeves, polished/clean shoes, neatly tucked-in shirt, spotless, & steamed clothing.

- Stay away from accessorizing your look too much (it may not look work-appropriate).

- Any visible nails (fingers/feet) should be properly trimmed, should look presentable.

- The breath should be odorless. Do not chew gum during the interview process. If you had chewed a gum right before the interview, rinse your mouth or drink some water not to smell strong.

- Do not spray strong fragrance cologne in excessive quantity.

- Well, obvious but often missed: put your phone on silent mode or switch it off.

- If you are wearing a digital watch, make sure it is silent and does not make any notification sounds, while you are in the interview.

- If you get stressed/tensed before an interview, do things that comfort you down (e.g. quick-meditation). Talking to your friends, family, mentor right before an interview also boosts confidence. You should present yourself confidently and professionally.

- Greet the interviewer/panel with a firm handshake and appropriate greeting.

- Carry along 3 hard-copies (no fold, no wrinkles) of your resume and keep some samples of your work or work-accomplishments, in case you get asked to present.

- For the first five minutes, you will have the maximum attention from the interview panel, make sure you practice well the 'About Yourself' question.

- Clarify any interviewer's questions/concerns about your profile in detail, if they have any.

- Practice the STAR (Situation, Task, Action, and Result) technique of answering situation-based questions. Provide examples from previous employment experience and fine-tune to bring forward an answer which resonates with your style and is believable.

- It is advisable to link your situation/story to some sort of deficiency that you had witnessed while in the situation. You will have all buy-ins from the recruitment panel when you answer in a way that elaborates on 'how you make decisions.' This helps the hiring manager recognize your core decision-making mechanism, leadership skills, resourcefulness in accumulating information, and your ability to make sound decisions.

- Make sure your speech does not make you sound brash about your achievements and convey others as a failure. If you must talk about a negative incident, talk with humility, and sincerity in your speech (it does not sound professional when you snitch about your co-workers and portray yourself as the only righteous person).

- Learn all the skills, keywords, and abbreviations mentioned in your resume. It can count against you if you are unable to explain anything from your resume.

- Do not interrupt the flow of the interview by asking questions before you are given a chance to ask. The interviewer will give you enough time to ask questions that you may have prepared from before/thought about during or after the interview.

- This book takes a holistic approach in providing Project Management interview questions and answers. It is recommended to read through the sample questions and answers a couple of times. Prepare the first few interviews by practicing your answers in front of a mirror or friends and family.

- Pay attention to your facial expressions and body language throughout the interview. Maintain eye contact with all the panel members.

- Research the company and interviewers' profile. Also, prepare your answer, 'why do you want to work for this company?'

- Do some research well before the interview regarding the salary range for the experience you possess, the industry you would be interviewed for and the job requirements.

- Do not initiate salary/compensation discussion, unless initiated by the interviewer. Always have a desired salary range in mind for each position (e.g., $65,000-$75,000 annually). If it is a contractual position, provide an hourly rate range for the compensation expectation.

- Some sample questions that you can put forth for the interviewer (it is wise to prepare a few questions ahead of the interview):

    a. Can you discuss more about the

project?
   b. What is the team-size?
   c. Methodologies or processes being followed
   d. Further steps in the hiring process
   e. How is performance measured in this role?

- Send a 'Thank You' note on the same/next day of the interview, expressing your gratitude for taking the time to interview you. Show how you are still excited to be a part of the team. Make sure the note is short and sweet. You can follow-up with the hiring contact after a week or so.

- Any answer should not exceed two to three minutes. You may need to adjust the answer time, depending on the question. Do not speak too fast or too slow. Watch your pace, specifically when the interviewers are making notes of your answers.

- As the interview gets wrapped up, exit while thanking the panel with a solid handshake. Express how you await a decision.

- At last, it is not always necessary that you get selected for the position you interviewed for, even though your interview went well. Do not feel demotivated or lose hope if you fail the interview(s). Remember, it is a process, and it may take time, and revisions. You should keep putting in efforts, and hard work in the right direction, and eventually, you will achieve success.

# First Few Days as a Project Manager

This section explains the day to day procedures of starting a new role; it also defines the key initiatives taken by Project Managers in the first few days. This daily routine may help you understand the Project Manager role even more clearly.

It is advisable for all Project Managers not to undermine this section. All information in this section is of high importance for a new or experienced Project Manager. Find below the practical and real-time process.

# Day 1

- I arrived at the workplace as early as 15-20 minutes before my reporting time. I wore regular office attire and was not super dressed up as this was going to be a regular workday.

- Before my start, I had received an instructive e-mail from the HR manager, which detailed reporting time and whom to contact along with some information on parking at the facility. I was also informed to bring some sort of identification card cards and void cheques for the payroll set up.

- At 8:45, I checked-in at the reception and asked for the concerned person. At security, I was given a visitor's tag as a means of identification for the day; which I was expected to return by the end of the day.

- Approximately at 9 AM, I met with the HR

Manager. After exchanging greetings, I was given a quick tour of the building and took that as an opportunity to introduce myself to the project personnel.

- After engaging in pleasantries with my new colleagues, there was an orientation program scheduled in one of the board rooms, for new hires, including myself

- I was provided with an overview of the company's operations, which included familiarization around its market position, customer base, accountability to reporting bodies, and other functional aspects.

- The presentation was well put together and a valuable source of information as I learned many key things about the company I will be working for. The Director of the department I was going to work for did a brief stint in explaining what they do, their strategic clients, processes that they have in place, active projects, and their existence in the specific business domain.

- Around noon, we had lunch, and I was accompanied by my Director.

- After lunch, I was shown a presentation on workplace safety, security, and insurance, which was later followed by a quick chat with HR on the internal Code of Ethics & Policies & Procedures.

- Orientation was concluded by the IT dept. where they briefed us on the company's IT tools, systems/software in use, and training on Intranet.

- I was provided with a work laptop and given a short walkthrough on accessing relevant files and drives.

## Day 2

- I submitted the required paperwork to the HR and also had my permanent access card handed over.

- With help from the department's Administrative Assistant, I set up my desk phone, desktop screens, and installed drivers for printing.

- (It is not necessary to know all the technical aspects on how to connect and get access to various devices to start up your work. The Support Team is always there to help you. Keep the support contact emails handy as you may come across some technical forest-up issues in the first few days.)

- Initially, I copied/CCed, my Director in communications initiated by me.

- I had to request the IT team to grant me access to the tool I was assigned to work on. They provided me with access and a quick walkthrough of the tool. The usability seemed a bit complex as this was my very first time using the particular tool, but with the help of training manuals, I picked up on the software after a few attempts/playing around.

- Later on, I attended a team meeting where project objectives, milestones, timelines, and my role in the project was discussed.

- I was proactive in the session and asked several questions related to the project to get the facts and figures straight.

- It was agreed that I'd meet the top management daily basis for a few weeks to discuss many deliverables/tasks on the project. Concurrently, similar meetings were to be established with my team so that I can further my understanding of the project and get status details.

- I was diligent in taking meeting minutes, and

identifying follow-up actions, which later did help me prioritize the tasks, and stay on top of my commitments.

## Day 3

- I checked my meetings for the day and responded to emails first thing in the morning. Later on, I continued to work on drafting my work management plan to ensure I capture the details of tasks on hand, deliverables, stakeholders, and cost and schedule needs of the project.

- A meeting was scheduled with my team today, and I was able to learn more about the project status and what tasks are team members assigned to.

- Towards the end of the day, I shared a high-level status update document with top management, which tracked the tasks assigned to me and their status to completion. I also forwarded working documents for their review.

## Day 4

- As I interacted with my team and my upper management team, I started working on making these meetings more organized and more meaningful by identifying relevant stakeholders and agenda for each meeting.

- I also spent time updating the RACI matrix, which I had started preparing the previous day.

- In addition to that, I spent my afternoon working on understanding the cost and schedule milestones of the project and where the project stands right now against these milestones.

- Working with a team member allowed me to

learn about their expertise and their allocated tasks to manage them efficiently.

- Gradually, I started obtaining positive feedback from my team and Director on the work I have been doing in the last few days. This boosted my confidence and inspired me to continue working smart in the achievement of project goals.

## Day 5

- After understanding the project details, I set up a meeting with the customer to ensure they are also aware of where the project stands and if they have any feedback.

- I exported the requirements from our requirements database and started building a compliance matrix to demonstrate the requirements we are compliant with and the requirements we are still working on to achieve.

- (By day 5, work seems to be getting simpler as my familiarity rises).

- In the meeting with the customer, I went through the requirements we have already implemented and gained their agreement and informed them about the path forward.

- For the next few days, I will be working to finalize, gather, and document new requirements with the stakeholders. Upon completing this, I intended to share the document with the stakeholders to establish the same level of understanding, so there are no disputes as the project progresses.

## Key Takeaways

The first few days as a PM in any domain may seem overwhelming because you are processing a lot of information; also trying to understand and retain most of the aspects of the project. This gets much easier later on once you acquire all the knowledge required and start utilizing it efficiently. The more you understand the nuances of the project, the easier it becomes eventually. Use your Project Management skills by making the project management process enjoyable.

- After working for a few months, you will be accustomed to the project management processes and it will become natural after some time.

- It is advisable not to get too excited in the beginning by overcommitting and later stressing yourself out to deliver on those tasks. Your efforts should concentrate on understanding the details, facts, 'As-Is' scenarios at best in the first few days and raising questions along the way as things get confusing. Don't make assumptions. Ask, Ask, Ask!!!

- Initially, obtaining clarity on the project is crucial than trying to narrow down the low-level requirements. In other words, if the bigger picture of the project is unclear/tainted, you may not get the requirements right later on.

- Brush up your knowledge of everyday tools. You always have to have a decent hold on basic applications as there likely be selected training for them and asking too many/inappropriate questions about their usage is just not very professional. Rather watch some online videos or read about them and polish your familiarity. It's okay to ask questions when you have tried to help yourself but did not succeed. Do the homework first, and later ask for help.

- Try to divide/organize your queries/questions by departmental heads.

- Technical questions should be directed to the technical team members and project-related questions, such as those on deliverables, timeline, and objectives can be asked to the project team.

- Requirements related questions should be clarified with users or stakeholders.

- Initially, you will not be connected with all the stakeholders; document all your questions and concerns clearly and clarify them once the meeting is set-up with the relevant stakeholders.

*That would be it PMs...*

*Hope you had a wonderful time reading this guide, a review on Amazon is always appreciated!*

*Wishing you success in everything you do. Good luck.*

Printed in Great Britain
by Amazon